6/99

WITHDRAWN

DISCOVERY READERS

Telephones, Televisions, and Toilets

How They Work—and What Can Go Wrong

Hello!

By Melvin and Gilda Berger

Illustrated by Don Madden

Ideals Children's Books • Nashville, Tennessee

The authors, artist, and publisher wish to thank the following for their invaluable advice and instruction for this book:

Jane Hyman, B.S., M.Ed. (Reading), M.Ed. (Special Needs), Ed.D. (candidate)

Rose Feinberg, B.S., M.Ed. (Elementary Education), Ed.D. (Reading and Language Arts)

R.L. 2.1 Spache

Text copyright © 1993 by Melvin and Gilda Berger

Illustration copyright © 1993 by Don Madden

Published by Ideals Publishing Corporation
Nashville, Tennessee 37214

Printed and bound in the United States of America.

Library of Congress Cataloging-in-Publication Data

Berger, Melvin.
    Telephones, televisions, and toilets: how they work and what can go wrong/by Melvin and Gilda Berger; illustrated by Don Madden.
       p.   cm. — (Discovery readers)
    Summary: A simple explanation of how three household items work.
    ISBN 0-8249-8645-8 (lib. bdg.)—ISBN 0-8249-8608-3 (pbk.)
    1. Household appliances—Juvenile literature. [1. Household appliances.] I. Berger, Gilda. II. Madden, Don, 1927- ill. III. Title. IV. Series.
TX298.B43   1993
643'.6—dc20
                                        92-18198
                                           CIP
                                           AC

Discovery Readers is a registered trademark of Ideals Publishing Corporation.

Produced by Barish International, New York.

Telephones, televisions, and toilets.
You use them every day.
Do you know how they work?

Telephones
You pick up the telephone.
You dial your friend's number.

Your friend answers.
You say "hello."
Your friend hears your voice.

How does the telephone carry your
   voice?
How does your friend
   hear you?

Hold your hand on your throat.
Say, "hello."

Do you feel something shaking
 inside?
Your vocal cords are shaking.
They are sending out sound waves.

Sound waves move quickly.
They zip through the air.
First they squeeze the air.
Then they let the air spread out.
Squeeze/s—p—r—e—a—d,
squeeze/s—p—r—e—a—d,
squeeze/s—p—r—e—a—d.

Now say "hello" into a telephone.
You send out the same sound
   waves.

Some go into a part of the
   telephone called the mouthpiece.

Inside the mouthpiece is a
    diaphragm (DYE–uh–fram).
The diaphragm is a thin, round,
    metal disk.
Electricity always flows through
    the diaphragm.
It comes from big batteries in the
    telephone office.

Sound waves hit the diaphragm.
The diaphragm bends forward.
A BIG burst of electricity flows out.

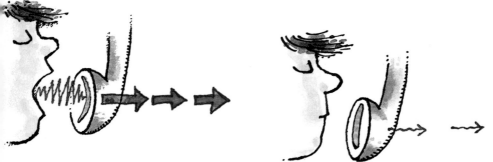

Then the diaphragm springs back.
Now only a small burst of electricity
    flows out.

Out go the bursts of electricity.
BIG/small, BIG/small, BIG/small.

They zip
   —from the mouthpiece into the
      telephone wire
   —through the wire to the
      telephone office
   —from the office to another wire
   —through that wire to your
      friend's telephone.

The bursts enter your friend's
　　telephone.
They go to a part of the phone called
　　the earpiece.

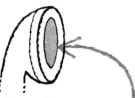

The earpiece also has a diaphragm.
The bursts of electricity make this
　　diaphragm shake.

 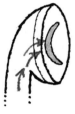

Big burst of electricity—diaphragm
　　bends forward.
Small burst of electricity—
　　diaphragm springs back.
Back and forth the diaphragm shakes.
The shaking diaphragm makes
　　sound waves.

You know these sound waves.
They are the sound waves for the
    word "hello."
The sound waves come out of the
    earpiece.
They pass through the little holes.
They strike your friend's ear.
Your friend hears you say "hello."

Some phones work in a special way.
Cordless phones work partly by
 wires and partly by radio waves.
The wires carry the bursts of
 electricity to the base of the
 cordless phone.
Radio waves carry the electricity to
 the phone's earpiece.

I found the lost cow, Grandma!

Cellular telephones are like
cordless telephones.
They use radio waves too.
Cellular telephones are good for
calling from a car.

The newest type of telephone is the
videophone.
The videophone is part telephone
and part television.
Videophones let you see and hear
your caller.

Telephones almost always work
  well.
But sometimes they don't.

Get a wrong number?
Dial again very carefully.

Your friend can't hear you?
Speak more loudly and right into the
  mouthpiece.

No dial tone?

Check that the phone wire is
    plugged into the wall.

Telephones reach everywhere.
You can speak to your friend.
You can speak to people all over the
    world!

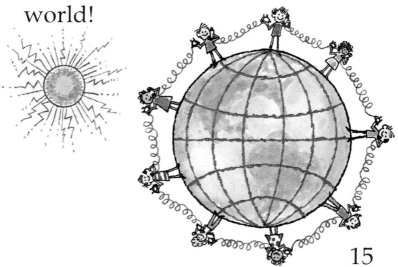

Televisions

Telephones send sound a long way.
But televisions do more.
They send sound and pictures a
    long way.

How do sound and pictures get to
    your TV set?
It all starts with a television camera.
The television camera works like
    your eye.

16

You look at a tree.
Sun shines on the tree.
The light bounces back off the tree.

Some of the light enters your eye.
You see the tree.

18

Let's say a television camera looks
    at a clown.
Light shines on the clown.
The light bounces off the clown.
Some of the light enters the
    television camera.
The camera "sees" the clown.

Inside the camera are three tubes.
One for red light.
One for blue light.
One for green light.

Painters get all the colors they need
    by mixing red, blue, and yellow.
Television gets all the colors it needs
    by mixing red, blue, and green.

The clown's suit has many colors.
Light from the red part goes to the
red tube.
Light from the blue part goes to the
blue tube.
Light from the green part goes to
the green tube.

What about the other colors?

All other colors are mixtures.
Their light goes to two or three
     tubes.

Yellow light is made up of red and
     green light.
Light from the yellow part splits.
The yellow light goes to the red and
     green tubes.

White light is made up of all three
   colors.
Light from the white part splits.
The white goes to all three tubes.

At the end of each tube is a screen.
A picture of the clown forms on
   the screen.
Each picture is in one color—red,
   blue, or green.

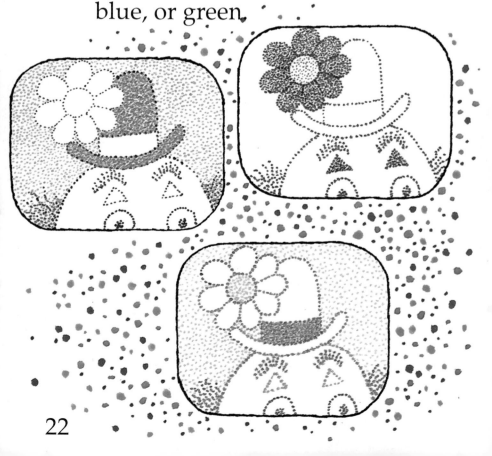

The screen is made of thousands of tiny spots.

The light from the clown hits the spots.

Each spot sends out a burst of electricity.

Spots with bright lights send out BIG bursts.

Spots with dim lights send out little bursts.

Spots with no light send out no electricity.

The electricity bursts copy the clown picture.

Black and white TV cameras are
    different.
They have only one tube and screen.
They send out a black and white
    clown picture.

Meanwhile a microphone "hears"
    the clown.
The microphone is like the
    telephone mouthpiece.
Sound waves enter the microphone.
They hit a diaphragm inside the
    microphone.

The sound waves shake the
diaphragm.
The shaking sends out bursts of
electricity.
The electrical bursts copy the clown
sounds.

Electricity from the camera.
Electricity from the microphone.
They join together.
This makes the television signal.

The television signal passes
into a wire.
The wire goes to the top of
a tall tower.
It is an antenna.
It is called the transmitter.

The transmitter sends out the
television signal.
It goes into the air.
The signal spreads out in all
directions.

Your television set may have an
    antenna.
It is called a receiver antenna.
The receiver antenna can be
    —on the roof of your house
    —on top of the television set
    —inside the set.

The receiver antenna picks up the
    television signal.
It gets it from the air.
It passes the signal into your TV set.

Some television sets use
a special antenna.
It is a big, round, metal dish.
The dish antenna faces the sky.
High in the sky is a satellite.
The TV station beams the signal up
to the satellite.
The satellite sends the signal back
down to earth.
The dish antenna picks up the
signal from the satellite.
It passes the signal into your TV set.

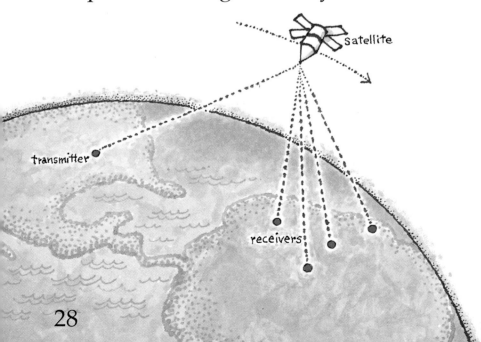

satellite

transmitter

receivers

Maybe your television set is
    different.
It does not have an antenna at all.
It is hooked up to a cable.
The cable is a very long wire.
It starts at the television studio.
It goes to peoples' homes.
The cable brings the television signal
    into your TV set.

satellite

Sometimes the cable company uses
   a dish antenna.
The company picks up the TV signal
   from a satellite.
Then it sends the signal through
   wires to each person's TV set.

Every TV set has a picture tube.
The picture tube works like a
 camera.
Except it works backwards!
The camera "sees" the pictures.

It changes the picture into bursts of
 electricity.
The picture tube receives bursts of
 electricity.
It changes the bursts back into
 pictures.

The picture tube mixes
 the three different colors.
Or it shows the picture in black and
 white.

Every television set has a speaker.
The speaker works just like the
  microphone.
Except it works backwards!
The microphone "hears" the sound.
It changes the sound into bursts of
  electricity.

The speaker receives bursts of
  electricity.
It changes the bursts back into
  sound.

That's how you see and hear TV in
  your house!

Is the picture too dark on your TV?
Adjust the *brightness* control.

Do the colors look washed out?
Adjust the *color* control.

33

Are the skin shades too greenish or
reddish?
Adjust the *tint* control.

Are you missing the picture and
sound?
Make sure your set is plugged in
and turned on.

Just as I suspected!

Good pictures and sound make TV
fun to watch!

Toilets

Every bathroom has a toilet.
But few of us know how toilets
work.

Most toilets have two parts.
The back part is the tank.

The front part is the bowl.
Both parts contain water.

The tank has a handle outside.

Inside is a rubber stopper.

Inside is also a round float.

The stopper covers an opening in
the bottom.

The float rests on top of the water.

To flush the toilet, you press the
handle.
This pulls up a metal chain.
The chain raises the rubber stopper.
The water in the tank flows out.
It passes through the opening.
It rushes into the bowl.

waste
pipe

The water swirls round and round.
The flowing water cleans the bowl.
It leaves through the bottom of the
   bowl.
The water then goes into the waste
   pipe.

The waste pipe is under the bowl.
It carries the water outside.

In cities, the water goes into a sewer
   pipe.
The sewer pipe carries the water to a
   treatment plant.
Here the wastes are taken out of the
   water.
The clean water flows into a lake,
   river, or ocean.

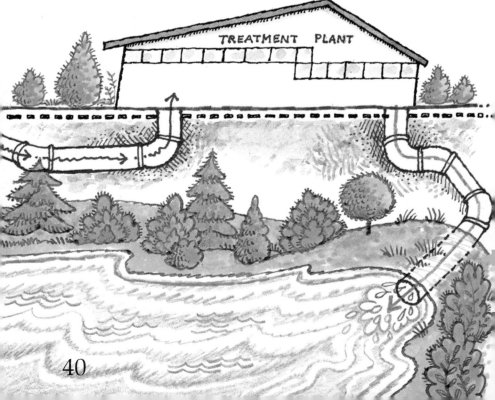

In the country, the water goes into a
 septic tank.

The septic tank is a big tank in the
 ground.
Germs in the tank take the waste
 out of the water.
The clean water then flows out of
 the tank.
The water mixes with other water
 under the ground.

Soon all the water is out of the
toilet tank.
Then two things happen.
The rubber stopper falls.
It covers the opening in the bottom
of the tank.
No more water can flow out.

The float also drops.
It goes to the bottom of the tank.
This lets fresh water flow into the
tank.
It is the same water that comes from
a faucet.

The level of water rises.
It raises the float.

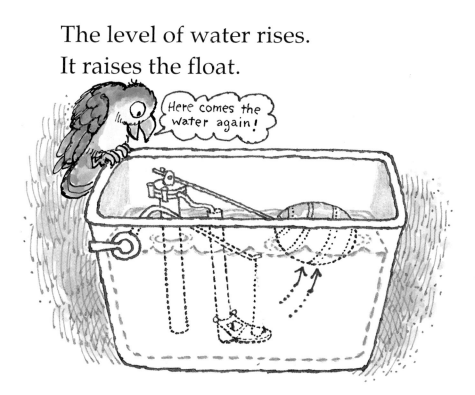

Soon it is up high.
The float shuts off the water.
Now the tank is full of water.
The toilet is ready to flush again.

What can go wrong with a toilet?
People sometimes toss an object
　　into the bowl.
The object gets stuck in the waste
　　pipe.
The toilet does not flush.

A plunger may help.
The plunger is a stick with a rubber
cup at the end.

Place the cup over the opening in
the toilet bowl.
Push down fast and hard on the
stick.
This presses the water down.
It may force the object out through
the waste pipe.

Sometimes the plunger does not
work.
Then ask an adult to try.
The adult may even have to call a
plumber.

Telephones, televisions, and toilets.

You use them every day.
Now you know
　　—how they work
　　—why they sometimes don't
　　　work
　　—what you can do to fix them.

Bye-bye

Index